THE IRON WORKHORSE

American Gas Tractors and Steam Traction Engines

Photography by Dave Arnold

Text by C. H. Wendel

Motorbooks International
Publishers & Wholesalers Inc
Osceola, Wisconsin 54020, USA

First published in 1988 by Motorbooks International Publishers & Wholesalers Inc, P O Box 2, 729 Prospect Avenue, Osceola, WI 54020 USA

Printed and bound in Hong Kong

The information in this book is true and complete to the best of our knowledge. All recommendations are made without any guarantee on the part of the author or publisher, who also disclaim any liability incurred in connection with the use of this data or specific details

We recognize that some words, model names and designations, for example, mentioned herein are the property of various manufacturers. We use them for identification purposes only. This is not an official publication

Library of Congress
Cataloging-in-Publication Data
Arnold, Dave.
  The Iron Workhorse.
  1. Steam-engines--Pictorial works.
2. Traction-engines--Pictorial works.
I. Wendel, C. H. (Charles H.)
II. Title.
TJ469.A76 1988   629.2'25   88-9317
ISBN 0-87938-314-3

Motorbooks International books are also available at discounts in bulk quantity for industrial or sales-promotional use. For details write to Special Sales Manager at the Publisher's address

To Timothy James, my second born. Thanks for your serenity and companionship.

On the front cover: *Of all early tractors, there were probably none more popular than the famous Rumely OilPull models. The unique OilPull design used oil as an engine coolant. By this means, plus a somewhat higher compression ratio than usual, Rumely was able to bill the OilPull as being capable of "burning all low grade fuels under any load." The huge box at the front of the tractor contains the cooling sections of the radiator system.*

On the frontispiece: *A well-used steam gauge on a threshing machine built by J. I. Case of Racine, Wisconsin.*

On the title page: *Taming the prairies was delayed until the arrival of the big steam and gas tractors. Although there were probably enough horses, mules and oxen to get the job done, there simply were not enough able-bodied people to handle the huge twenty-horse hitches required for the job. Photographs of the time illustrate anywhere from twenty to forty horses or mules pulling a single large combine. Using these big draft hitches required several men on hand at all times.*

On this page: *In the early 1900s, most major steam engine builders offered at least one compound engine design, and that included this 1902 Russell sixteen-horsepower tandem compound. Built at Massillon, Ohio, the Russell was constructed along very conservative lines in what today might be termed a no-frills engine.*

On the back cover: *The famous OilPull trademark remains today as one of the best known manifestations of a unique, yet very popular, tractor. One secret of the OilPull design was the use of an oil coolant instead of water. This provided the higher cylinder temperatures required for successful use of low-grade fuels such as kerosene.*

Contents

Introduction

Although invention of the steam engine is generally credited to James Watt, American efforts at steam engine building are almost as old as the republic itself. Initially, Americans were content with the huge and relatively inefficient engines of the day. By the 1830s, however, the concept of adapting steam engines to self-propelled vehicles was gaining some credibility, and was in fact, instrumental in development of the steam locomotive and a nationwide network of railroads.

Immense problems followed every single step of steam engine development. While it was possible to pour excellent castings by 1850, the ability to machine these castings was severely limited by the simple lack of machine tools or any uniformity of machining practice. As is often the case with inventive minds, concepts far outdistanced technology; steam engine development actually went hand in hand with developing mechanical technology, with each separate industry goading the other to higher achievements.

Threshing machines first made their appearance in the 1830s, but the McCormick reaper of the period soon made it possible to harvest grain on a larger scale than ever before imagined. The flail was totally inadequate to harvest the grain, and hand-powered or horse-powered threshers were likewise incapable of the task. Thus the growth of the steam engine as applied to farm power went hand in hand—and sometimes led—the development of American agriculture. As technology improved so did the available motive power; as power sources improved, the size of the crop increased, and the new bounty required more power—a full circle of agriculture and industry needing each other.

Early steam engines were of portable design, capable of being moved from place to place by

A first step toward mechanical power on the farm was the portable steam engine, similar to the J. I. Case model illustrated here. A few farmers used treadmills or sweep powers so that animal power could be converted into useful work, but these were the exception rather than the rule. Instead, most farmers continued to do all their work by hand, much as it had been done since the dawn of time.

horses. Surprisingly, very few skid-mounted or stationary steam engines were used on American farms. The portable engines led to semi-portables with a crude driving mechanism from the engine to the rear wheels, but requiring a team and driver to steer the engine. Full steering became available in the 1870s.

Steam pressures in the 1870s ran in the sixty to eighty pounds per square inch range, and rarely past 100 pounds. Improved steel and better boiler designs eventually led to steam pressures of 200 pounds in some cases, with 125 to 150 pounds being common usage.

Steam engineers faced innumerable problems. Dirty water in the boiler might cause foaming or priming, both conditions tending to carry water into the engine cylinder along with the steam. This dangerous condition was likely to cause breakage of a cylinder head or other serious damage to the engine. Boiler water was often loaded with lime and other minerals, all of which remained in the boiler to burn and crust on the boiler plate and tubes. Boiler compounds

The next logical step in the evolution of the steam engine was the traction engine. It was designed so that it could move itself, but it still depended on a team and driver to do the steering—and perhaps lend extra pull to get through the tough spots. Note the big cast-iron driver's seat mounted alongside the boiler.

were not widely used, but some engineers recommended "dumping a hatful of potatoes" into the boiler every morning. Apparently the starch of potatoes was beneficial in minimizing the attachment of lime to the boiler.

Fuel for the boiler was another constant problem. Poor quality coal was full of soot that plugged the flues. Soft wood had little heating value and required almost continual firing. Wet hardwood lumber, such as sawmill slabs, could make a very hot fire indeed, but learning to fire with sawmill slabs was, at the least, an educational experience.

Country roads were usually rough at best and impassable at worst. Sandy pockets or sink holes could spell immense problems; dropping a big steam engine into one of these spots could halt progress for several days, or at least until enough men, horses and other equipment were amassed to extricate the engine from its predicament.

Bad bridges were one of the most frequent sources of concern, and numerous engineers went to their untimely end when their engine fell through an unsafe bridge. More often than not, serious bridge accidents caused horrible injury to the engineer as a result of scalding from broken steamlines damaged in the accident. Farm magazines, threshermen's associations and other groups lobbied government officials intensively to repair or replace poor bridges.

The steam tractor occupies an important niche in the history of American agriculture, making its appearance in the 1890s. These early models used a stationary engine mounted on a chassis, usually a steam engine chassis. The result was originally called a gasoline traction engine, with the term "tractor" finally becoming accepted usage.

There was no lack of difficulty with early tractors. If all was well, starting a huge two-cylinder engine with a ten-inch bore and twelve-inch stroke was reasonably simple. Should any of the three requirements—electricity, fuel and compression—be lacking, it was a certainty that if the engine started at all, it would only be achieved with great difficulty. There was no such thing as a "starter" motor. The only starter around was the operator on the end of a four- or five-foot crank.

The business end of this J. I. Case steamer illustrates the simple cylinder design. In steam engine jargon, a simple cylinder is one in which the steam is expanded but once and then exhausted into the atmosphere. Tandem compound and cross compound engines turn boiler steam into a smaller, high-pressure cylinder. After expansion and useful work there, it travels to a larger low-pressure cylinder for further expansion, and finally to the exhaust. This Case engine is of single-cylinder design, although some makers preferred a double-cylinder style, technically, a double simple engine.

Interestingly, tractor development gained immensely from the parallel research of automobile engineering. The auto industry was quick to adapt the new, lightweight and high-speed designs of European practice, and by about 1915 a few farm tractors were following this concept, at least to some degree. The 1920s saw these ideas placed into practice with the Farmall tractor of International Harvester.

Steel wheels, on the other hand, prevailed for farm tractors until the 1930s. When Allis-Chalmers first introduced the rubber-tired tractor, critics within the industry flatly stated that the steel-wheeled tractor would always retain a place in American agriculture. The fact was that by 1940 the vast majority of American farm tractors used rubber tires.

Farm machinery development closely followed the perfection of other machines and other phases of mechanized agriculture. Manufacturing technology in the early twentieth century called for farm machines built largely of wood or cast iron. Little was known of all-steel construction, much less the high-strength alloys which began making inroads during the 1930s. The wood thresher was one of the first all-wood machines to disappear. Its construction, and occasional dust explosions within the machine as a result of static electricity, could cause disastrous fires. The all-steel thresher of the early 1900s eliminated this difficulty.

Looking at today's farm tractors and equipment, it seems impossible that our mechanical origins began with the steam engines, the huge tractors and the great threshers of the past. Despite our disbelief, these are our roots. Those that remain for us to see and enjoy are truly a great heritage. More importantly, they remain as a lasting tribute to our sturdy ancestors who simply got tired of doing things the way they had always been done, and put their shoulder to the wheel in search of a better way. Had they not succeeded, we might still be cutting grain with a scythe and threshing it with a flail.

A J. I. Case gentle giant, complete with Case's bald eagle trademark in relief on the front of the boiler.

Chapter 1

Manufacturers

Advance-Rumely and Gaar-Scott

Very few Gaar-Scott tractors were built, so it is more than a little surprising that any have survived their working life and the junkman's hammer. Gaar-Scott was bought out by Rumely in 1911, although the firm had just developed a prototype of this fascinating tractor a year before. The trademark of the Gaar-Scott Tiger line of engines and threshers naturally illustrated a tiger astride both earthly hemispheres.

Gaar-Scott Company of Richmond, Indiana, entered the tractor business only months before being bought out by the M. Rumely Company in 1911. Rumely had just announced their OilPull, and probably saw Gaar-Scott emerging as a major competitor. Rumely built a few TigerPull tractors, at least until the parts inventory was exhausted.

Although Advance-Rumely Thresher Company was purchased by Allis-Chalmers Manufacturing Company in 1931, the world-famous Rumely OilPull tractor is still one of the most popular vintage tractors ever built even now, over half a century later. Perhaps it was the unusual design, or perhaps it was the unique booming sound of the exhaust that continues to draw collectors and spectators alike to these fine examples of early tractor engineering.

RUMELY
OilPull
TRACTOR
LA PORTE, IND.
REG. U.S. PAT. OFF.

GUARANTEED
TO BURN SUCCESSFULLY ALL GRADES OF KERO-
SENE UNDER ALL CONDITIONS, AT ALL LOADS
UP TO ITS RATED BRAKE HORSE POWER.

Avery

Avery Company at Peoria, Illinois, used a tubular radiator on its early tractors. In this instance, the engine exhaust terminated in vertically disposed nozzles within the top section of the radiator. The stronger the exhaust, the stronger the induced draft and the greater the cooling effect on the water tubes. Similar systems were employed by Rumely, Hart-Parr and several other builders.

A close-up of the valves and rocker arms of an Avery tractor gives an idea of their immense size. Due to the great depth of the cylinder head, valve stems of ten inches or more in length were not uncommon. The slow-speed engines were of relatively high displacement, and this required valves and passages of a large diameter. Slow-speed engines also had the distinct advantage of a long life. Despite what might now be called a crude or cumbersome design, the many engine designs of the past have been valuable building blocks that have brought us to the present state of the art; today's engines will probably look awkward a century from now as well.

Engines like the Avery used this unique steering system. The steering wheel shaft terminated at the large bevel gears turning a pair of spur gears running to the steering shaft. This shaft is a screw with a very fast pitch. Riding on this screw is a huge nut to which is attached the steering arm. Although this system gives the appearance of being complicated, it represented an important forward step away from the chain-and-bolster steering that characterized early designs.

This Avery tractor dates from the 1920s and represents a substantial improvement due to the adoption of a cellular radiator in place of the earlier induced-draft, tubular design. These Avery models actually shifted the entire engine forward and back on the tractor frame to engage the proper gears for forward and reverse travel.

Early gas tractors were usually equipped with a starting handle or some other device intended to ease the task of bringing a big piston of six inches or more in diameter past dead center. Ignition would take place as the engine passed dead center. This was usually aided by priming the engine, accomplished in turn by opening petcocks in the top of the cylinder and squirting in a charge of raw gasoline from an oil can. The petcocks also served as a compression release, and after starting were closed. This Avery tractor has a special starting device: a long handle which engaged notches in the rim of the flywheel.

This close-up of the Avery tractor radiator shows the engine exhaust at the very top. Nozzles inside the stack point upward, and the engine exhaust induces air currents to flow up and around the vertical water tubes. Water circulates through the myriad tubes with assistance from a substantial centrifugal pump mounted on the tractor engine.

Next page
Avery Company went bankrupt in 1924 and reorganized as Avery Power Machinery Company. Although the latter firm continued building a few big tractors like this one, the 1920s saw the beginning of the end for them. Farmers were looking toward the new, lightweight, row-crop designs, easily adaptable to virtually any crop or tillage practice.

Buffalo-Pitts

*Relatively rare today is the
Buffalo-Pitts steamer, even though
a substantial number were built.
Buffalo-Pitts had a long and
illustrious history going back all
the way to the 1830s when the Pitts
brothers, John A. and Hiram,
invented the first "groundhog"
thresher, the immediate
predecessor of the huge grain
threshers of the early twentieth
century.*

28

J. I. Case

Smoke billows from the stack on this big 110 horsepower Case steamer as it raises steam for another day's work. J. I. Case even built a few gigantic 150 horsepower steamers about 1907. Ostensibly intended as road locomotives, these tremendously large engines had no peer for size and power. Unfortunately, none of these engines still exist.

Previous page
Engines like this 110 horsepower Case, the 140 horsepower Rumely or any one of several other makes, were instrumental in doing what had never been done before— turning vast expanses of the plains into productive cropland. A centuries-old growth of anything from impenetrable brush to tangled prairie grass had resisted conventional efforts of converting the prairie to cropland until the arrival of the steam tractor.

A few engines like this 110 horsepower Case actually used a power steering system—and it was full power, too. The engineer steered the engine with a lever which engaged clutches to turn the steering shaft left or right, depending on where he wanted to go. This system was imperative on an engine of this size; to have steered the engine by hand would have required herculean strength.

The Case Eagle trademark is probably one of the best known in the farm equipment industry. J. I. Case began business in the 1840s as a thresher builder, with the record books now showing that Case built more threshers than anyone else. Case also built more steamers than anyone else, despite the fact that the Case engine was often criticized by the competition. Looking back on the Case odyssey leaves no doubt that Case was

determined to have satisfied customers. The story is often told of how J. I. Case himself was once called to a farm where the firm had sold a thresher. After unsuccessfully working with the machine in order to get it set, Case ordered that it be pulled off to the side, whereupon he doused it with kerosene, set it afire and sent a telegram that a new machine should be shipped forthwith to the farmer.

J. I. Case Company offered an extensive line of cross-motor tractors during the 1920s. As is evident here, the engine was situated transversely over the tractor frame. A cellular radiator is obvious; this great improvement was probably borrowed from automotive engineers. The notches on the inside rim of the flywheel engaged the latch of the starting crank.

Hart-Parr

The huge Hart-Parr 30-60 model was largely responsible for putting this Charles City, Iowa, builder on the map. These big two-cylinder tractors were shipped to the wheat producing areas by the hundreds, being used to pull giant plows which turned over the prairie sod. The 30-60 Hart-Parr was also quite popular for road-building work.

Hart-Parr claimed to be the "Founders of the Tractor Industry," and truly deserve the accolade: its plant in Charles City, Iowa, was the very first one in America devoted solely to the production of tractors. The firm's 28-50 of the 1920s was an extremely powerful four-cylinder model.

International

International Harvester Company gained almost immediate fame with their huge Titan tractors, such as this sixty horsepower model. Its big two-cylinder engine was cooled by pumping the jacket water through the big rectangular radiator at the front of the tractor. Many of these huge tractor radiators carried one hundred or more gallons of water. That was an awful lot when it had to be carried from the pump to the radiator with a couple of buckets!

The big International Harvester Company trademark was featured on the front of the water tank for the Titan 10-20 tractors. Built in the late 1910s, the 10-20 Titan did battle with the Fordson tractor of Henry Ford. The years from 1915 to 1925 have been called the period of the great tractor wars due to the intense competition for a piece of the market pie. Survival of the fittest was often the rule as tractor manufacturers by the hundreds went bankrupt in the 1920s.

Minneapolis
Threshing

The Twin City 40-65 tractor was
fairly popular among
threshermen, and a substantial
number was purchased by county
governments for road-building
projects. The huge tank at the front
of the tractor is actually a radiator
with a capacity of more than 100
gallons. Cooling fans at the rear of
the radiator pulled cool air
through the tubes.

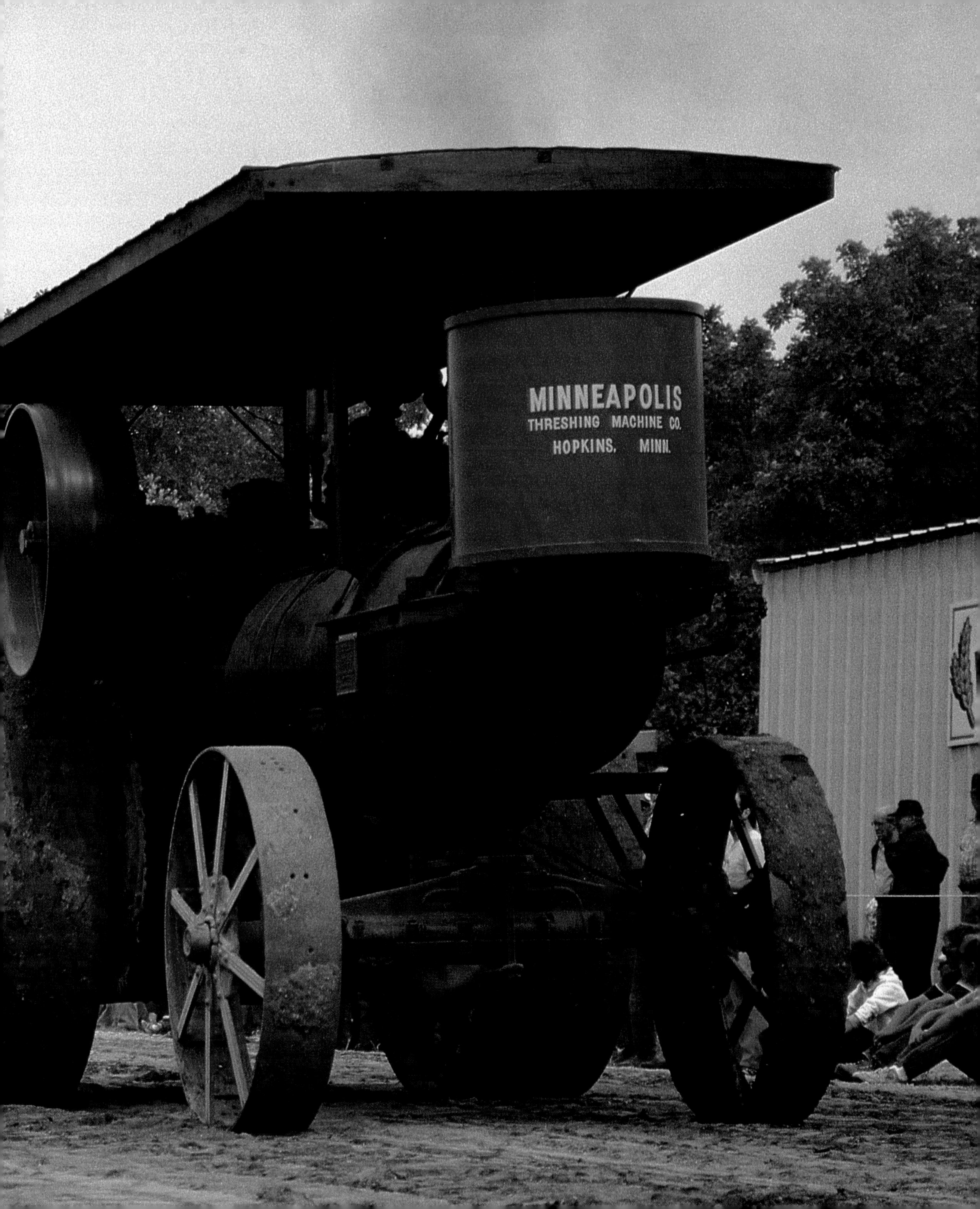
MINNEAPOLIS
THRESHING MACHINE CO.
HOPKINS. MINN.

Previous page
Minneapolis Threshing Machine Company of Hopkins, Minnesota, was a well-known steam engine and thresher builder—one of their engines is shown here making its way through a parade. Curiously, no one manufacturer dominated the field, although certain geographical regions seemed to prefer one make to another. Threshermen epitomized individuality and independence, each one going with the engine and thresher he was convinced would do the best job for the least cost.

A conventional cellular radiator is evident on the Minneapolis tractor built by the threshing machine manufacturer of that name from Hopkins, Minnesota. Eventually, tractor designers began borrowing more and more from the advances made in automobile engineering. This included everything from automotive steering systems to high-speed automotive engines, cellular radiators and many other design features.

Huge tractors powered with gasoline engines had become prominent by 1910. For a few years at least, more power meant more iron, and tractor sizes seemed to have no limits. This gigantic 60-90 Twin City tractor was built by Minneapolis Steel & Machinery Company of Minneapolis, Minnesota. This fourteen-ton behemoth carried a six-cylinder engine. The fuel tank held ninety-five gallons!

Pioneer

Pioneer tractors were built in Winona, Minnesota. Chances are that very few were built, and of these, even fewer survive. Designed for large power requirements, the Pioneer—like the other heavyweight models—was totally unsuited to the average 160 acre midwestern livestock farm. In fact, once the prairies were tamed by plowing the virgin sod, the jumbo steam tractors had largely served their tour of duty, except for the running of large threshers or similar jobs.

48

This surviving example of a
Pioneer tractor was ordered by
Edgar and Harry Knox of Alpena,
South Dakota, in 1910. For this
tractor they payed $2,850, plus
another $75 freight from Winona,
Minnesota. A new engine was
installed in 1917 for the sum of
$900. About 1924 it was used for
breaking some sod, and after that
it sat out in the elements for fifty-
five years. During this half
century, many of the steel parts
were utilized for other purposes.
Fortunately, this old tractor has
been revived and can now be
looked on as a nostalgic reminder
of the past.

50

Chapter 2

Components

Although most steam engines were top-mounted—that is, the engine was mounted up top of the boiler— a few, like this Avery engine, were of under-mounted design. This concept had the distinct advantage of eliminating any mechanical strains on the boiler plate, leaving the boiler to achieve the single goal of making plenty of steam for the double-cylinder engine. This engine was built in Peoria, Illinois, by Avery Company.

Previous page
Advance-Rumely engines enjoyed an enviable reputation for fine workmanship and construction. This double butt-strap boiler is probably suitable for operating pressures of 150 to 190 per square inch, depending on the thickness of the shell. Double butt-strap boilers are characterized by a heavy splice plate on the inside and on the outside of the boiler shell at the joint of the seam. Row after row of heavy steel rivets gave great strength to this method of construction.

Boiler designs were almost as numerous as the varying engine configurations. The Huber was characterized by a return flue boiler, a design that carried the fire in a main flue with a diameter of twenty-four inches or larger. The combustion gases then returned through numerous small flues, ending up in the smokebox above the fire door, and finally up the smokestack.

Jumbo engines were fairly popular in the midwestern states, and although these engines followed the major features on several other lines, they were highly regarded for their quality construction. They were built by Harrison Machine Works of Belleville, Illinois.

Previous page
Side-mounted steam engines have stub axles for the drivers mounted directly to the side of the boiler. Rear-mounted engines used a single axle mounted just behind the boiler. Simple cast-iron spur gears were used in most cases. Each method of building had its followers, and in the end, virtually all of the well-known designs were fairly successful.

Steam engine horsepower ratings were, and still are, a source of confusion. An engine with a nominal rating of sixteen horsepower would likely have an output of approximately fifty brake horsepower. Most companies used the nominal rating, but Case engines used the brake horsepower rating. Thus, their fifty horsepower engine probably had a nominal rating of perhaps fifteen or sixteen horsepower. The nominal rating appears to have had roots in the amount of power delivered by a single horse—horsepower, if you please.

Now a rarity, the New Giant engine was built by Northwest Thresher Company of Stillwater, Minnesota. Organized about 1874, Northwest was a successor to the earlier Minnesota Thresher Manufacturing Company, also of Stillwater. The big red tank carried a reserve cache of boiler feed water.

Every steam engine was equipped with a governor. The brightly colored governor balls of this Judson governor are mounted on spring steel leaves; a suitable linkage moved a valve within the governor in response to speed changes. Judson, Pickering, Gardner and other firms specialized in governor building. The typical unit, when properly adjusted, was very responsive to load changes. The leather drive belt is missing from this governor and has probably been removed by the owner lest it get wet in a sudden thunderstorm.

With steam pressure at slightly over seventy-five pounds per square inch, this Case engine is nearly ready for business. Most engines operated at 100 pounds or more, since steam at lower pressures carries too much moisture and washes away protective lubricant from the cylinder walls. In its heyday, this engine probably operated at 150 pounds, and a capable engineer could maintain this pressure within a close range by adjusting the draft or adding feed water at just the right time.

By 1910, tractor companies abounded. For reasons now unknown, much of the inventive activity in the tractor business seemed to be centered in Minneapolis, Minnesota. Such was the case with this gigantic seventy horsepower tractor from Diamond Iron Works. The cooling radiator is situated between the big drive wheels.

A most important part of any traction engine was the safety valve. In the good old days, however, constant popping of the safety valve was not considered to be the mark of a good engineer. Also evident on this engine is the steam whistle. It could be heard for several miles, and most threshermen used a code of long and short blasts to send messages ranging from, "It's time to start," to "We need the water wagon right away."

Next page

Gas Traction Company of Minneapolis, Minnesota, made headlines in about 1909 with their Big 4 tractors. These huge four-cylinder machines were of comparable size and output to the 30-60 Aultman-Taylor or the 40-65 Twin City tractors of the period. The tubular radiators were built of lightweight sheet metal for maximum cooling effect. They were not, as has been supposed, made from old tubular steam boilers.

One of the most impressive tractors ever built was that of the Nichols & Shepard Company. Fortunately, at least this huge 35-70 model still exists. The gigantic radiator sits forward like a masthead; above it are two large fuel tanks, plus a small tank for the gasoline used in starting the engine. When all engine components were in top condition, starting one of these giants was usually accomplished with relative ease, but if something, anything, was awry— well, that's another story!

69

Previous page
Aultman-Taylor tractors were built in several models, all following this general pattern of a huge radiator mounted ahead of a four-cylinder cross-mounted engine. Placing the engine in this fashion eliminated the need for bevel gear drives and permitted the use of relatively simple cast-iron spur gears to deliver power from engine to drive wheels. Top engine speed was usually under 500 revolutions per minute, and a typical model might have an engine with a seven-inch bore and a ten-inch stroke.

Aultman-Taylor tractors like this big 30-60 model used two clutches—the one on the flywheel was used for forward travel and the one on the opposite end of the crank was used for reversing and to operate the belt pulley. The belt from the flywheel is connected to a large centrifugal cooling water pump mounted beneath the tractor frame. One of the two cooling fans is also evident.

72

The belt pulley side of the 30-60 Aultman-Taylor tractor illustrates the large clutch within the belt pulley itself. This clutch operated the belt pulley, and to reverse the ground travel it was necessary to stop the belt pulley and engage a large idler gear. This done, the tractor could then be reversed. The big drive wheels on this tractor are ninety inches in diameter.

75

Many early tractors, including this 1915 Nichols & Shepard model, used extremely complicated carburetor and manifold systems. One reason for this was that the use of low-grade fuels, such as kerosene, made it imperative to design the intake manifold for a certain amount of preheating. Achieving this rather elusive goal brought countless designs and configurations, few of which were completely effective.

Restoration of a vintage engine or tractor often involves considerable expense. The engine cylinder shown here has had extensive welding work on the jacket. This work often requires the service of an expert welder, in addition to many hours of preparation before the actual repair process can begin. Restoring an engine from this condition to good-as-new is a challenge that requires dedication and work, the reward coming on that day when the engine is under its own power again.

77

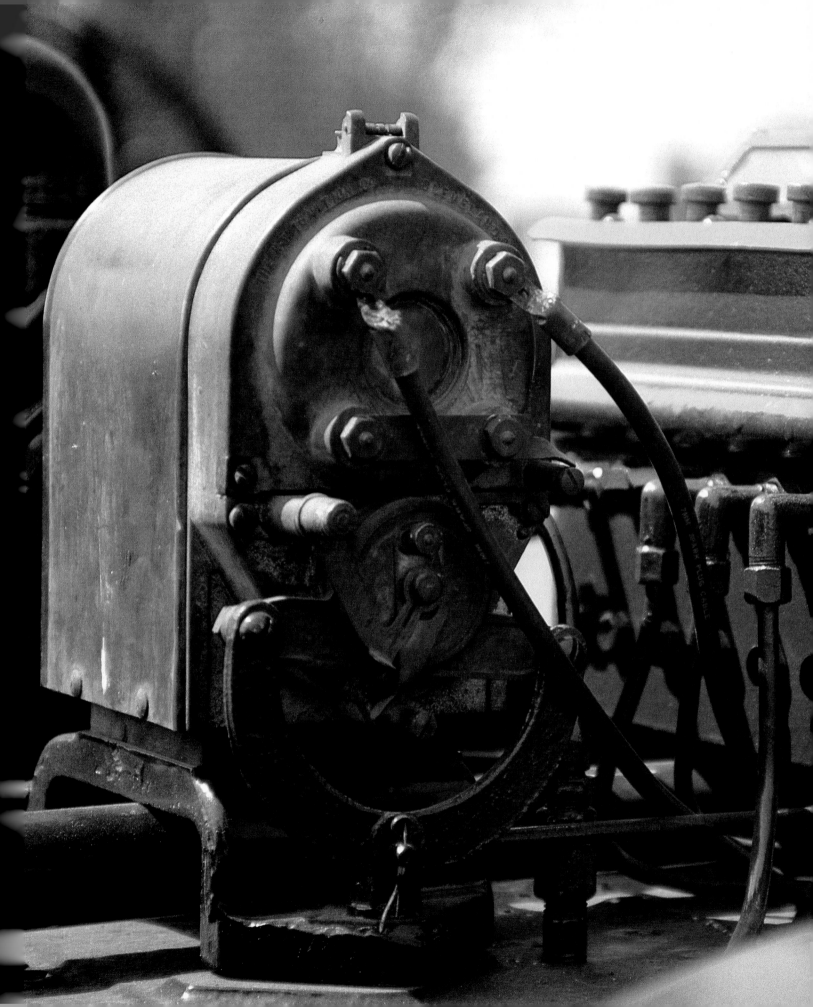

The magneto was really the
heartbeat of an engine; if it had a
weak or intermittent spark,
everyone knew about it. The
magneto shown here was made by
K-W Ignition Company and had a
reputation among some folks as
being the best on the market.
Dozens of other fine magnetos were
available, including Bosch, Dixie,
Simms, Eisemann and Wico.

Force-feed lubricators were present on most early tractor engines. The lubrication system consisted of a combination force-feed and splash arrangement. Some points, such as the engine cylinders, were lubricated by the force-feed method while the connecting rod bearings were often equipped with little dippers which picked up the oil from cast-iron splash pans. The pans were fed with tubes from the lubricator. Shown here is a Madison-Kipp lubricator; other well-known makes included Manzel and McCord.

Early steering mechanisms
usually consisted of a stiff axle
mounted on a bolster and moved
one way or the other by means of
heavy chains attached to a
windlass. The worm gear
terminated at its top end with the
steering wheel. This system
required considerable strength on
the part of the engineer and also
carried a certain amount of risk
in jerking the wheel out of the
engineer's hands should a front
wheel suddenly drop into a hole or
strike an unseen obstruction.

Walking up to a gigantic rear wheel of the 110 horsepower Case steamer provides some perspective of its immense size. Ample quantities of cast iron were used in the construction of these engines as cast iron was relatively cheap and plentiful. Eventually the trend would move toward high-strength alloy steels which provided the necessary support with far less weight.

Duties

Frequent cleaning of the boiler tubes was required for efficient use of fuel. Even a thin coating of soot reduced the boiler's steaming capacity. Flue cleaners took on many shapes, but all were mounted on a long rod which the engineer laboriously pushed through one tube after another. Large engines might have well over a hundred tubes, each being perhaps as long as ten feet. Cleaning the flues was just another of the unglamorous duties required of the steam engineer.

85

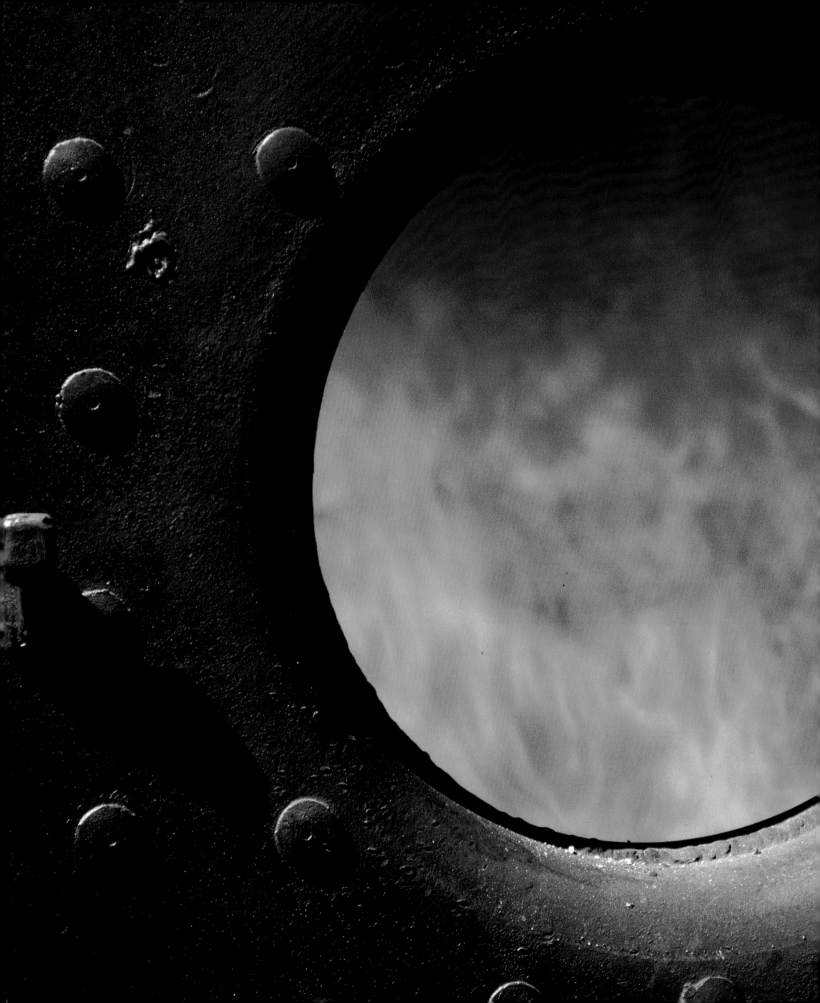

The open fire door shows the nearly white heat in the firebox. Engineers were careful to maintain a fairly thin and "clean" fire—one that was very hot, but with a minimum of smoke to deposit soot in the flues. A heavy draft was induced by the steam exhausting into a vertical nozzle placed within the smokestack. This made the draft quite responsive to the engine load. The steam engineer was always careful to keep the fire door closed as much as possible to prevent cool air from entering and causing unequal strains on the firebox metal.

A firebox is at the opposite end of these boiler tubes. The smokebox also contains the exhaust nozzle, and it can be seen here pointing upward into the smokestack. This created an induced draft of intensity varying directly with the power requirements. Each boiler tube is carefully rolled into the tube sheet. Here, a couple of tubes to the lower right appear to be leaking slightly. When the boiler is hot, however, the leaking will probably stop.

Steam engineers were in general agreement that the ordinary slide valve which distributed steam to the two ends of the engine cylinder had several disadvantages. The chief problem was that the pressure in the steam chest also pushed the valve very tightly against its seat, adding an immense amount of friction, and wasting power. The Gould Balance Valve Company of Kellogg, Iowa, offered their so-called balanced valve which was a substantial improvement in design.

Chapter 4

Special equipment

Many steam engine builders were able to supply their engines with the special equipment for road roller duty, as is this J. I. Case model. Road building was an important duty for steam traction engines—after all, rural America of the early 1900s had almost no macadam roads, much less any paved highways.

91

THE RUSSELL & COMPANY
BUILDERs
MASSILLON OHIO
SANDY & DOUG LANGENBACH

RUSSELL

Previous page
The big canopy on steam traction engines was not necessarily provided for the comfort of the operator, but was intended to protect the engine and boiler from the elements. Many threshermen had no shed large enough to house the steamer, so it spent 365 days per year in the weather.

Engines used for plowing and other drawbar work were often equipped with special extension rims that provided greater flotation and additional traction. Most steamers had a top road speed of only two to three miles per hour, but the immense gear reduction from crankshaft to the final drives provided the torque power needed to pull gigantic plows and other equipment.

The so-called Baker fan was first used by Abner D. Baker who built steam traction engines in Swanton, Ohio. This fan with its four paddles is capable of working virtually any engine built, since the horsepower requirement increases substantially with increasing shaft speed. Each of the four paddles is twenty-four square inches. Baker fans are often used at engine shows as a method of providing a sufficient load to present a demonstration of vintage power at work.

A lone sentinel on the prairie, this drive wheel is believed to be from a 25-50 Minneapolis Threshing Machine Company tractor built between 1910 and 1915. It weighs approximately 1,500 pounds, stands eighty-eight inches tall and is twenty-four inches wide. Owner Fred Bruns, from Hecla, South Dakota, found this wheel, its mate and four other large drivers on a common shaft being used to pack soil.

Steam traction engines converted water to steam and steam into useable power. After its one-way trip through the engine, the steam was exhausted into the atmosphere. Boiler feed water was required in copious quantities; hence the water wagon, whether of steel or wood construction, was an imperative accessory.

Next page
The good old days included big water wagons like this one to carry feed water to the steam engine. Sometimes a creek was nearby, making it easy for the water boy to keep pace filling the tank. Occasionally it was far enough away that the water boy's task became a man-sized job. Depending on the size of the engine, two or three of these large loads of water were required for each day of operation.

Previous page
Fourteen-bottom plows like this one were frequently used in taming the prairies of the United States and parts of Canada. These huge plows were hand operated, with the attendant working on a huge running board just ahead of the individual plow levers. Despite the nostalgia connected with this work, it was in fact very strenuous and required great stamina in addition to an inordinate amount of skill.

This demonstration of a John Deere plow behind a J. I. Case steamer demonstrates not only the plow in action, but also shows the large running board or platform on which the plowman worked. In actual practice, one or two men were required for this task, plus another as engineer and still another person as the fireman. Support included one or more water wagons, plus other teams and wagons with coal or other fuel.

Previous page
*This relic of the past is folded up
and ready for the next job. Huge
threshers made it commercially
feasible to plant and harvest vast
areas of wheat and other crops.
This machine is of wood
construction but uses an all-steel
feeder. Thresher factories employed
thousands of highly skilled
carpenters to build these machines.
Many were immigrants who were
able to ply skills learned over an
almost interminable
apprenticeship period.*

*One of the better known thresher
lines was the Red River Special
built by Nichols & Shepard
Company of Battle Creek,
Michigan. These bright red
machines covered the American
prairies by the thousands at one
time. Nichols & Shepard used a
special thresher design that
incorporated what they termed,
"The Man Behind the Gun."*

Standing silently by the roadside, this thresher now serves as the backdrop for a signboard advertising a South Dakota threshing show. Were it sufficiently animate to talk, this old thresher, like thousands of others, could probably relate some fascinating tales of the good old days.

110

Chapter 5

Shows

Numerous steam engine shows held each summer serve to once again bring vintage engines to life. A home-built test unit shown here is putting this engine through its paces. Steam traction engines under load have a most pleasant sound, especially to steam lovers; they often refer to it as "stack music."

Previous page
The combination of coal smoke and steam cylinder oil provide a whimsical picture of the past. These engines have all undergone painstaking, and sometimes expensive, restoration to like-new condition. Now that their working days are over, these giants remind us of our agricultural past.

Aboard a steam traction engine once again, this seasoned engineer beams with joy. In the heyday of steam, local communities accorded a certain prestige to those talented individuals who had risen to the rank of steam engineer. Learning the fine art of firing the boiler was by itself a task that could only be gained by experience. Handling the throttle of a big steam engine is a thrill unlike any other in the human experience.

116

This gigantic 110 horsepower Case had a nominal drawbar rating of about thirty-two horsepower. Fully equipped with extension rims on the rear wheels, it pauses here for a crew change on the plows during a demonstration run. In actual operation, the black smoke billowing from the stack was not looked upon as good practice, since this tended to coat the boiler flues with soot, and soot lowered the heating efficiency tremendously.

119

Previous page
With the setting sun, these steam-powered prime movers signal that day is done. In usual practice, engineers would bank their fires in the evening so that the boiler would still be warm by the next morning. Then, after cleaning the flues and shaking the grates, all was ready for a new fire to greet a new day of threshing.

This big Twin City 40-65 tractor now finds its primary function is to serve as the flag bearer in a parade. Huge, majestic tractors like this one weighed from twelve to fifteen tons—and occasionally even more. The Twin City 40-65 used a big four-cylinder engine set in line with the frame. Hand cranking was the only way to start the engine, so Twin City engineers mounted the cranking accessories on the back of the tractor transmission where they engaged a live shaft back to the engine. Thus, this tractor was cranked directly from the operator's platform.

Previous page
A line-up of early farm tractors at an engine show attracts many spectators. Beyond the nostalgia and reminiscences of yesteryear, these assembled pieces of mechanical art represent the collective work of thousands, all hoping to make life a little easier.

Nichols & Shepard Company of Battle Creek, Michigan, was one of the early steam engine and thresher builders. This company was a partner in the great 1929 merger which formed Oliver Farm Equipment Company. Fortunately, a fair number of Nichols & Shepard engines still exist; this one is belted up to a thresher for a demonstration run.

Now used at a steam engine show, this Prony brake probably began life with a steam engine or tractor builder for determining the exact horsepower output. The Prony brake consists of a revolving wheel within a large band. The outer band is connected to a platform scale, and by squeezing the outer band against the revolving wheel, the precise horsepower output can be determined on the basis of rotative speed and the force on the scale.